Secrets of the Snow

Visual Clues to Avalanche and Ski Conditions

Edward R. LaChapelle

UNIVERSITY OF WASHINGTON PRESS *Seattle*

INTERNATIONAL GLACIOLOGICAL SOCIETY *Cambridge, England*

GREYSTONE BOOKS *Vancouver/Toronto*

Secrets of the Snow

Visual Clues to Avalanche and Ski Conditions

01 02 03 04 05 5 4 3 2 1

Printed in Canada by Friesens Corporation, Altona, Manitoba
Text design by Veronica Seyd
Cover design by Val Speidel

University of Washington Press
PO Box 50096, Seattle WA 98145-5096

Greystone Books, a division of Douglas & McIntyre Ltd.
2323 Quebec Street, Suite 201, Vancouver, B.C. V5T 4S7

All photographs are by the author.

The paper used in this publication is acid-free. It meets the minimum requirements of American National Standard for Information Sciences—Permanence of Paper for Printed Library Materials, ANSI Z39.48–1984.

This book is dedicated to
Dolores LaChapelle and David LaChapelle,
who shared those years when we all were learning
secrets of the snow.

Contents

Preface

In 1969 I completed a book called *Field Guide to Snow Crystals*, published by the University of Washington Press. This book subsequently went out of print in the 1980s, but was revived as a reprint edition in 1992 by the International Glaciological Society and is still available today. The *Field Guide* told only half of the story, that of snow in the microscopic world. The present volume tells the other half, snow in the macroscopic world, where visible features of the snow surface delineate the external forces that have shaped it. In the same size and format as the *Field Guide*, *Secrets of the Snow* is intended as a companion volume to that book.

The concept of this present book originated many years ago, when I worked on snow and avalanche studies for the U.S. Forest Service at the Utah ski area of Alta. One sunny winter morning, following a fall of fresh snow, my wife and I were seated at the breakfast table together with a visitor. Our view out the window took in many slopes of the ski area. She and I were discussing which ski runs would have the best snow, where skiing would be poor, as well as the general character of snow deposited by the previous night's storm. Our visitor, puzzled at this conversation, asked, "But how can you sit here and know what the snow is going to be like?" We knew because experience had taught us to recognize from the visual appearance of the snow surface the whole story of its deposition and probable skiing character.

During much of my professional career as a snow and ice scientist, I photographed snow features wherever I found them. About twenty years ago I compiled and annotated many of these photos as part of a larger work on snow and avalanches that for many reasons never saw the light of day. The photos and accompanying text gathered dust in my files until they turned up recently, suggesting that the long-delayed second half of the snow story could now be told. With the text and photo collection revised and enlarged, the result is *Secrets of the Snow*.

Edward R. LaChapelle
McCarthy, Alaska
2001

Secrets of the Snow

Visual Clues to Avalanche and Ski Conditions

Introduction

This book is about snow in the mountains. Whenever snow lies on steep slopes, it may become unstable and slide away as an avalanche. Hence there will be frequent references to snow stability and clues to interpreting it. But this book is not about avalanches per se; texts dealing specifically with avalanches are listed in the bibliography. This book is intended for the interested layperson, for ski tourers, for avalanche workers, and for alpine snow country dwellers everywhere. The experienced ski or avalanche professional will find much here that is familiar (and perhaps a few things that are not). For these people, the photographs may help focus attention on aspects of snow that have become an unconscious part of their snow skills, and this introduction to "snow secrets" may help form a sound basis for developing those same skills. For all readers of whatever interest, the important theme here is to develop the ability of paying attention to the many visual clues to snow behavior.

The snow surface is a sensitive register of the forces that mold it, exhibiting a constantly changing picture whose details bear a large body of information about snow stability and behavior. The molding forces are the flow of thermal and mechanical energy at the earth's surface and the stress of gravity. These forces leave their footprints in the form of clues to snow behavior and stability for those who are alert to looking for them and have the understanding to interpret them. The primary clues are visual, but the kinesthetic sense plays an important role, and even auditory signals offer systematic evidence. This book summarizes many of the visual clues. Although the other sensory signals are important, communicating their description for the most part lies beyond the capacity of the printed page.

The snow surface is so variable and ephemeral that it would be a tough job indeed to cover all the potential forms in a discussion such as this. Many major and typical features presented here are illustrated by the available photos, collected by the author during a half century of watching what snow is and does. This text, however, serves a purpose beyond exhibiting basic characteristics. It will emphasize looking at the snow with attentive eyes and mind and illustrate how visual perceptions, interpreted in the light of meteorology and snow physics, lead to deductions about snow trafficability, ski conditions, and avalanche formation.

This look at snow characteristics will begin with the general, large-scale features of snow on the landscape, then will proceed to details of snow shapes and textures seen on a smaller, more intimate scale. The general features are largely those of form, which often can be seen at a single glance taking in snow distribution across a landscape. The details of shape and texture, on the other hand, require selective perception of, and closer attention to, separate elements in the visual field. Both are essential for gaining insight into what a given body of snow is like and what it is apt to do next.

Large-Scale Features

To begin, compare figures 1, 2, 3, and 4. Each is a mid-winter scene of snow-covered alpine terrain, but startling differences exist among the overall appearances of these landscapes. The degree to which the snow covers the landscape varies widely, owing to wide differences in snow-cover depth and distribution. These basic differences stem from equally basic differences in climate. This is the starting point for evaluating ski conditions and snow stability: what kind of snow climate must be considered? The internal structure, distribution, likely size, and location of avalanches are all related to snow climate. The principal determinants of snow climate are the total amount of snowfall (both annually and by individual storms), the typical range of winter temperatures, the frequency of supercooled water clouds, the amount of wind, the altitude above sea level, and the relation to timberline. How these factors are combined sets the stage for the basic character of the snow cover, how it can bear loads, and the nature of its stability. The principal snow climate regimes of the western United States were first pointed out by Andre Roch in 1949 and since then have become a standard consideration in such diverse applications as designing ski areas and forecasting avalanches.

Figures 1 and 2 are characteristic of a maritime climate. Nothing is left visible in figure 1 except the white of rime-cemented snow. Wind, riming, and damp snow falling close to the melting point have all combined to plaster every exposed surface. In figure 2 most small-scale features of the terrain have disappeared under a smooth mantle of snow, suggesting snow depths that may run two or three

1 Olympic Mountains, Washington

2 North Cascades, Washington

3 *Wasatch Mountains, Utah*

4 *Front Range, Colorado*

meters or more. Frequent deep snowfalls, temperatures close to the melting point, a deep snow cover, and infrequent clear skies (limited surface radiation cooling) are the features of a maritime snow climate, which tends to produce a high-density, well-compacted body of snow. With snow surface temperatures often close to or at the melting point, good deep powder skiing usually is found only during or immediately after snowfalls. Once melt starts in fresh new snow, its skiing character rapidly deteriorates, earning such pejorative terms as "peanut butter," "gunk," or "Mount Hood powder." A maritime snow cover tends to make a stable base where avalanching often is confined to surface layers of new snow. Prolonged thaw or rain nevertheless at times can induce avalanching by lubricating a deeply buried ice layer.

The typical snow climate depicted here is not guaranteed in even a maritime climate, for occasionally an unstable snow structure may develop, especially in anomalous winters of shallow snow that doesn't blanket the landscape. This principle applies to other climate zones as well, for instance when exceptionally deep snowfalls bring a maritime character to inland areas. A snow cover like that in figures 1 and 2 typically is stable in its deeper layers, but this always has to be checked against such additional evidence as winter snowfall history.

Figure 3 shows at a glance an inland snow cover with intermediate depths. A substantial blanket of snow perhaps one to two meters deep covers the landscape, but it does not clothe completely the steeper ridges and rock faces. Because they are much less plastered with rime-cemented snow, the trees have readily shed most of their snow load with a day or two of sun following the last snowstorm. (More evidence to be gleaned from the way snow lies on trees will be treated later.) This is a scene typical of the intermountain ranges of the western United States, where substantial precipitation falls most winters, and higher altitudes and clearer skies assure a colder snow-cover regime than is found in coastal climates. Wind action probably has been moderate here. Local cornices and drifts can be seen, but there is no evidence of wholesale rearrangement of snow across the landscape. The colder snow regime promises more consistently good skiing conditions. A stable base for the snow cover cannot be taken for granted in such a climate, for the degree of depth hoar formation varies widely from year to year according to early winter weather patterns. The initial assumption from the evidence in figure 3 has to be that snow stability is uncertain. Additional evidence—from meteorological history of the winter, from internal snow structure, or from both—must be sought. External evidence alone does not give a clear answer about probable internal structure of the snow cover in this climate.

Figure 4 depicts a complete contrast to the winter scenes of figures 1 and 2. A quick glance at overall patterns shows that in figure 4 the snow obviously does not cover the landscape. A much smaller total snowfall has been extensively rearranged as the wind sweeps large areas of ridges and peaks free of snow. The continental climate and high altitude assure a cold temperature regime with dry and poorly consolidated snow except where it is packed hard by the wind. Individual snowfalls tend to be small in such climates, accumulating in small increments to form a poorly compacted snow cover. Frequent clear skies assure low snow surface temperatures, which, combined with shallow snow depths and strong temperature gradients, virtually guarantee extensive depth hoar formation. The initial assumption from the scene in figure 4 is a basically weak snow cover with depth hoar highly probable. In this case a second assumption is also in order. Avalanche-prone areas are going to be highly localized on lee slopes, unlike the situations in figures 1, 2, and 3, where avalanche release zones can be more widely distributed, often with less pronounced relation to wind direction. Based only on the landscape overview in these four photos, surface conditions in figure 4 focus attention on lee drift areas as likely avalanche paths. In figures 1, 2, and 3, the more extensive blanketing of the landscape also shelters from view the character and distribution of earlier snowfalls. The enveloping snow surface in these cases leaves uncertain what lies concealed underneath.

These four photos are chosen as clear type examples of three major snow climates. Intermediate gradations between these types obviously exist. As already mentioned, each locality can also experience a shift in snow climate from year to year or even alternate snow climates within a single season. It is not sufficient to know the geographical location; the observer also needs to know what snow climate is currently being expressed.

Wind on Snow

Like the sea, the snow surface is constantly shaped by the wind. Unlike the sea, snow carries the history of this shaping within its internal and external textures. Reading this history is a vital part of reading the snow surface, for wind deposition of snow usually governs both the skiing quality and the mechanical character of unstable slab layers prone to avalanching. For sophisticated forecasting of snow conditions, instrumentation like recording anemometers is essential to follow wind behavior. When such sensors are lacking, as is the usual case for a backcountry ski tour, the wind-influence history has to be observed subjectively, and its past effects deduced from the record etched in the snow surface.

The evidence of wind action may be obvious and current, as the snow banners show in figure 5. A single glance at such a situation yields several important pieces of information. A strong wind blows at the ridgetops where loose snow to windward is available for transport, but does not seem to be affecting the lower slopes. The location of lee slopes and thus drift deposition are clearly indicated by the prevailing wind direction. Absence of snow on trees in the foreground and middle distance suggests that at least a day or two of sun and/or wind has occurred since the last snowfall. The extended snow plumes on a clear, dry day are going to experience sublimation. This means that some of the transported snow mass will not return to earth, and that which does will consist of crystal fragments modified by sublimation as well as mechanical breakage. Such snowdrifts deposited on lee slopes will usually have a higher density and be harder than those deposited during snowfall and high humidity.

5 *San Juan Mountains, Colorado*

Highly localized slab avalanches are apt to form on lee slopes, their degree of stability depending on the rate of loading and the nature of the underlying surfaces, with danger often persisting a long time at high altitudes and low temperatures. Skiers may be challenged by breakable crust or boiler-plate slab. Caution is indicated above timberline until more information can be collected.

Local Wind Features

Figures 6, 7, and 8 illustrate the way wind influences can be interpreted by examination of local features. In figure 6, a layer of fresh snow exhibits a smooth, featureless surface except where broken by ski tracks in the foreground. The drifts are shaped by the building, an obstacle to the free flow of the wind. The sharp, clean edges of the drifts and roof deposits indicate drifting under moderate winds, as does the featureless character of the snow surface, which would have been etched with micro-relief by stronger wind. The small cornices on the roof ridge indicate that the prevailing storm winds were toward the camera, and the size and location of the extensive snowdrifts point to accumulation from several storms. Although the snow surface and appearance of ski tracks indicate a fresh snowfall, perhaps just ended as the cloudy sky begins to break, the tree shows no accumulation at all, suggesting it must have been swept free by the wind. The ski tracks reflect at least 15–20 cm of soft snow, possibly that of the inferred fresh fall. The overall deduction is that of an appreciable new snowfall accompanied by moderate winds, a situation likely to produce good powder skiing and potentially some soft slab avalanche formation. The degree of instability would have to be determined from other factors such as density of snow, character of the previous snow surface, or, more directly, test skiing to see if fracturing will propagate. The roof drifts on that building might be a safe place to begin such testing.

Figure 7 presents a more complex wind history. The size of the cornices in the center indicate that prevailing storm winds blew across the ridge from right to left, transporting snow from the wind-

6 *San Juan Mountains, Colorado*

ward timbered slope to the cornices and the lee slope below them. Recent winds have altered this pattern. First, a wind blew in the opposite direction from the prevailing one and transported an appreciable amount of snow. It could have accompanied snowfall or picked up previously deposited snow from the surface. This drifted snow formed eddies around the cornices, whose normally lee faces acted in this case as windward obstacles, leading here to

7 *Wasatch Mountains, Utah*

the formation of miniature "wind scoops." This transported snow has had a chance to age-harden—at least a few hours is a reasonable guess—and then a strong wind has blown upslope to erode this age-hardened surface and etch the final texture visible in the photo. These features suggest that prevailing lee slopes where slab avalanches might normally be found have been wind-scoured. On the other hand, the countervailing wind may have deposited slabs in unexpected places. In any case, ski conditions are going to be changeable from slope to slope.

Figure 8 speaks very clearly to wind action on snow, lots of wind. The scoured and scalloped surface is formed by high wind velocity, probably hurricane force at times. The original snow must have been deposited and reworked by a sequence of windstorms, ensuring age-hardening and hard, high-density snow. If new snow has recently fallen, it too has been blasted away by the wind, while the hardened old snow beneath remained. With no evidence of melt around the exposed grass patches, the inferred site and season can be high altitude, early to mid winter, and probably both. Scarcity of snow on the distant slopes points to very early winter. Elsewhere the snow cover will be very shallow, with depth hoar formation likely, and highly localized deep drifts may exist on lee slopes, with surface erosion by wind scour and sublimation. Very scattered, hard slab avalanches may be possible, and ski conditions certainly will be poor if any skiable surface can be found at all.

Figure 9 presents another complex situation. The snow obviously must be solid and hard, for the skier hardly makes a dent in the surface. The peculiar surrounding drifts are old ski tracks made earlier in a layer of new snow deposited on top of the hard surface. Later a strong wind completely removed this new snow except where it had age-hardened following compression under passing skis. If the hard surface also exists on nearby lee avalanche slopes, it may provide a good sliding surface for a slab consisting of at least two layers, the original new snow deposition with more recent wind-transported snow on top. As shown as well by the wind-blown character of the more distant slopes, skiing is going to encounter highly variable surfaces.

Ridgetops are obvious places to find snow revised by the wind, but not all ridgetops are the same. As with the broader character of terrain, snow climate also varies on this smaller scale. In figure 10,

8 *Front Range, Colorado*

moderate wind accompanying a snowfall has left hard-packed snow on exposed places and softer snow easily tracked by skis nearby. More distant snow surfaces suggest a mid-winter snow cover with occasional strong winds. This scene is typical of snow accumulation above timberline, with both ski conditions and slab avalanche development highly dependent on slope orientation.

9 *Wasatch Mountains, Utah*

Figure 11 is untypical. Here, very strong wind apparently has compacted a shallow snow layer into a surface like concrete, where even boots leave no mark. More than wind is at work here. The shallow, high-density snow at high altitude has developed hard depth hoar, a condition first identified by the Japanese snow scientist Akitaya. A strong temperature gradient working on the fine-grained, compact snow has produced tiny depth hoar crystals among the grains, cementing them together into an exceptionally hard struc-

10 *Mount Hutt, New Zealand*

11 *Mount Asahidake, Hokkaido*

ture quite the opposite of conventional, weak depth hoar. When Japanese scientists were developing techniques to prepare competition surfaces for the 1972 Winter Olympics in Sapporo, they invited their best ski racers to ski on natural surfaces and report which would be best for racecourses. The skiers chose hard depth hoar from sites like this, and the scientists went on to perfect ways to prepare matching artificially hardened snow for the competitions instead of the more conventional icy surfaces made by spraying with water.

From these few examples it should already be clear that the external appearance of snow as it lies on the landscape bears many clues to ski and avalanche conditions. The visual appearance alone seldom tells the whole story, but it often points in the direction to look for more clues. The photos here are discussed strictly on the visual evidence they present, but an observer in the field obviously will have access to other information as well. The current and recent weather usually are known; altitude, climate zone, and terrain aspect can be recognized; penetration by skis or boots tells a story; and recent avalanches may have been observed. The character of the snow begins to emerge as the visual evidence is continually supplemented from other sources.

As this review of wind effects continues in the next chapter, it will soon become apparent that this agent is responsible more than any other for textural details of the snow surface. Almost every pattern in this surface is related in some way to the wind, either directly by mechanical influence or indirectly by the role of wind in evaporation, condensation, and melting.

Small-Scale Wind Features

Figure 12 shifts to a fine detail of snow structure (scale given by ski poles). The many thin layers, each only a few millimeters thick, are formed as small variations in hardness engendered by fluctuations of wind velocity during deposition. Subsequently, part of the deposited snow has been eroded by a stronger wind that etched the exposed layer edges to reveal the hardness variations. This feature serves as a reminder of the layered structure of snow that can lead to the formation of failure planes. Part of a layer of new snow may

12 *Wasatch Mountains, Utah*

13 *Wasatch Mountains, Utah*

slide away as a soft slab with no obvious distinction from the rest of the new snow remaining behind, but in each case some minor fluctuation in snow deposition, perhaps a brief wind gust or interruption of snowfall, has left behind the seed of a failure plane and later slab avalanche release.

Figure 13 moves to a larger scale with evidence of thicker layering. Again, a snowfall has been deposited with wind and weather variations during the storm. Again, a subsequent wind has eroded this deposit. The steps in the eroded surface exhibit the original

14 *Wasatch Mountains, Utah*

deposition layers, which are especially clear in the upper part of the photo where the eroded surface intersects the layers at a steeper angle. This picture speaks of a crust breakable by skis and repeats the message about snow stability: new snow has fallen, then later been transported by the wind, hence lee slopes should be suspect.

Similar layering is seen in figure 14, where the photo covers an area about one meter wide. In this case the overhanging weight of a small cornice has caused the snow to deform so that the layers follow the bending cornice curve.

15 *Wasatch Mountains, Utah*

Layering also occurs on a larger scale. In figure 15, the thick depositions of successive snowstorms are visible in this large cornice. The added weight of each storm bends downward the previous layers until the lowermost one rests on the lee snow surface below. Continuity of these layers with snow or rock anchors to the windward gives substantial strength to the cornice and resists its fall.

Once a cornice does fall and leaves a fracture face, as in the foreground of figure 16 (where another cornice is in the act of falling in mid-distance), a different situation prevails. When a new cornice builds outward from the face left behind by the first cornice's fall, there is no continuity between old and new snow layers, leaving the new cornice poorly attached to the snow behind it. Second generation cornices are much more unstable than first generation ones. Beware!

16 *Bridger Range, Montana*

The etching of deposition layering (figs. 12, 13 and 14) is possible only in relatively hard snow that can resist complete removal by strong wind. Softer, looser snow seldom exhibits such layer etching because it is simply blown away, but it does develop other surface textures more closely related to behavior of the wind rather than the layering of the snow. Figure 17 shows a transition to softer snow (skiing quality improves!) where moderate wind effects have delineated some of the layering, but where surface rippling from wind eddies also begins to appear. This rippling is more pronounced in figure 18 (still softer snow) and becomes prominent in figure 19, where the surface texture is now dominated by the complex interaction of wind and snow, with crystals being sorted by size and minor density variations along the snow surface. The main evidence about snow conditions to be gleaned from this kind of surface is that the

17 *Wasatch Mountains, Utah*

fresh snow is relatively soft (it may harden with time) and the wind has been in the low to moderate range below 10 to 12 m/s (meters per second). Skiing may be good and soft slab formation possible if other factors are favorable for avalanching.

Before continuing to examine new snow surfaces, it is appropriate at this point to consider briefly the type of rippled pattern in figure 20. At first glance this resembles the scale and pattern of those in figures 18 and 19. In this case, however, the snow is old summer firn, and the small-scale surface pattern is produced by a strong (over 20 m/s), warm summer wind that generated rapid melt from heat transfer and water vapor condensation (the latter releases the large amount of heat used to generate the vapor from liquid water). The

***19** Wasatch Mountains, Utah*

rippling effect here comes from small-scale instabilities in the turbulent heat transfer by a strong, warm wind over high-density and mechanically stable snow. A rippled surface thus has to be interpreted in light of snow age, season, and temperature regime as well as physical appearance. Figure 20 also exhibits a large-scale channelled structure related to meltwater percolation in snow, a topic that will be discussed later in regard to figure 66 (p. 82).

20 *Olympic Mountains, Washington*

Following again the diminishing effects of wind on the snow surface, figure 21 illustrates new snow deposition accompanied by light wind of 4 to 6 m/s. A larger-scale surface pattern is visible in the form of subtle undulations seen by illumination that brings out the very shallow surface texture relief. The ski tracks suggest 30 cm or more of new snow has fallen, enough to generate both good powder skiing and the prospect of soft slab releases if a poor bond exists with the previous snow surface.

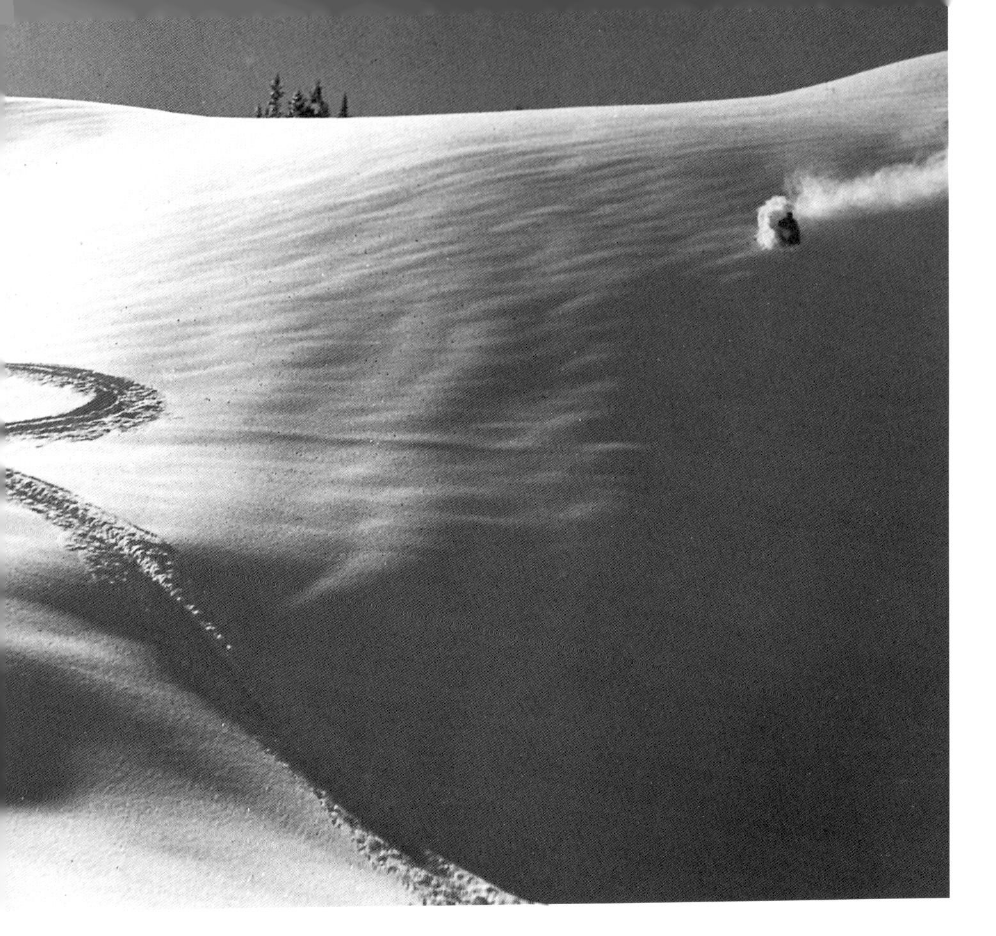

21 *Wasatch Mountains, Utah*

In figure 22, wind influence has diminished to negligible. The textured new snow surface reflects the granularity of clumps of individual snowflakes (aggregates of snow crystals) undisturbed by the wind. Figure 23 is a closer view of this same condition. Such snow offers good skiing and seldom forms soft slabs unless the bond to the underlying snow surface is extraordinarily poor. This might happen when a fragile surface hoar layer (see fig. 25) has been buried by the new snow. If the density is sufficiently low, 50 kg/m^3(kilograms

22 *Wasatch Mountains, Utah*

per cubic meter) or less, loose snow avalanches are possible, and sluffing on steep slopes is common. A note of caution is in order here. Wind-free deposits like this usually cause little avalanche danger in themselves, but even a shallow layer of such snow may conceal more dangerous conditions underneath. The surface texture never tells the whole story.

Figures 17, 18, 21, and 22 display additional evidence because they show skiers in motion. Experienced skiers will quickly recognize that

23 *Wasatch Mountains, Utah*

all of these photos, covering a spectrum of wind effects, represent good powder snow skiing as evidenced by the character of the tracks, the trailed plumes of snow dust, and the posture of the skiers. This kind of visual evidence tells something about the new snow depth and degree of cohesion. An observer actually skiing these slopes would also have available the kinesthetic evidence of ski reaction to the snow, or its "feel." Subtle differences in density and stiffness gradients in new snow can be felt through the skis. Good powder skiing can be found in a wide variety of new snow, some favoring soft slab formation and some not. Experienced persons develop a fine sense of discrimination between stable and unstable new snow just from the ways skis react. Clarifying this ability is a challenging area for research in psycho-rheology.

Although the kinesthetic discrimination of density and other mechanical properties related to snow stability are difficult to quantify, the quality of deep powder skiing depends very much on an identifiable density profile in new snow. Even in a deep layer of low-density snow, below 70 or 80 kg/m^3, skis find little flotation, and the skiing surface actually is the base underneath the new snow. Such skiing is photographically spectacular, with large rooster tails of snow dust, but does not provide the optimum sensation to the skier. The best skiing is found in a new surface snow layer with a density gradient increasing from top to bottom. In order to provide the best flotation, the lower layers should reach a density around 150 kg/m^3 while still retaining very low ski resistance. This usually occurs when a snowstorm begins with a significant coating of rime on the falling snow crystals. Ideally, the degrees of crystal riming and wind velocity diminish as the storm progress, leaving low-density fluff on top.

A peculiar form of surface snow texture develops when alkali dust contaminates a new snowfall during deposition (fig. 24), the so-called "salt storm" encountered in the intermountain ranges of the western United States. These storms usually occur in March, when the

24 *Wasatch Mountains, Utah*

low-altitude snow cover has disappeared from the alkali flats of the Great Basin region of Utah and Nevada. Storm winds then pick up the alkali dust and mix it with snowfall on the adjacent mountain ranges. This kind of snow has a mealy, lumpy texture when disturbed and exhibits remarkable powers of adhesion. For one thing, it adheres to ski running surfaces so efficiently that skiing becomes almost impossible. A skier can point skis into the fall line and walk straight down a steep slope. In the experience of this observer, it also never forms avalanches. When this kind of snow sticks to power-line insulators and then starts to melt, it can result in spectacular short circuits. The actual content of solids can run to many grams per cubic meter of snow.

Other Snow-Surface Features

The discussion of snow surface texture now moves to other features unrelated to the degree of wind action. These are features characteristic of winter snow and subfreezing temperatures. There is another whole domain of surface textures produced by various forms of summer ablation, but these largely relate to firn fields on glaciers and play little part in winter ski conditions or avalanche formation. Surface melt features and wet snow conditions will be discussed later.

Figure 25 depicts a close-up view of a snow surface spanning an area about 15 cm wide. The feathery crystals are surface hoar deposited by condensation of atmospheric water vapor directly as ice at subfreezing temperatures, without the presence of a liquid

25 *San Juan Mountains, Colorado*

26 *Wasatch Mountains, Utah*

phase (dew). They form a fragile, loose structure at the surface that severely inhibits bonding of a subsequent snowfall to the older snow underneath. A thick layer of surface hoar can in itself release as small, loose-snow sluffs, but these offer little danger. On a solid base, it can also provide very fast and pleasant skiing. Once buried beneath accumulating snow, however, surface hoar becomes one of the commonest causes of slab avalanche release by providing a very weak layer that encourages failure at the slab base. Such a buried layer is visible in figure 26, where a thin slab of snow cut vertically

27 *Wasatch Mountains, Utah*

from a snow cover is illuminated from behind by sunlight. Figure 27 gives a close-up view. The thin, bright layer across the slab is buried surface hoar, with a thin, dark band of ice crust, prominent in figure 27, just beneath it. This is an especially favorable combination for slab avalanche release. Evaluating such a hazard is tricky because surface hoar forms erratically across the landscape according to local surface cooling and variations in supply of atmospheric moisture. Once formed, it can be easily and erratically swept away by wind.

28 *Rocky Mountains, Colorado*

Figure 28 shows the fuzzy appearance of a snow surface coated with surface hoar. (No, the camera is not out of focus.) An earlier deposition of rime on the grass stalks has occurred from wind-blown, supercooled cloud droplets. A more complete discussion of surface hoar and rime, with microphotographs, is given in *Field Guide to Snow Crystals*, the companion volume to *Secrets of the Snow*.

In favorable light conditions, surface hoar can reflect many bright pinpoints of sunlight from crystal surfaces, which provides a sparkling appearance. In figure 29, however, the many bright points

29 *Mount Yotei, Hokkaido*

of specular reflection are from a different source, the surfaces of large, rime-free stellar crystals deposited by snowfall. Lack of rime on such crystals produces a new snow type that can form shallow, very unstable slabs under the right wind and temperature conditions. Such snow does not usually build up into deep and dangerous slabs unless the wind is strong enough to drift and fragment the crystals. The tendency to form unstable soft slabs of dangerous size

30 *Wasatch Mountains, Utah*

increases in new snow as the degree of riming on the crystals increases. The observer must be cautious, though, for the presence or absence of rime on crystals at the surface may not reflect what lies underneath.

The rough snow surface in figure 30 is produced by the complex interaction of sunlight, melt, and evaporation, which leads to enhancement of surface relief. The net result is to generate an increasingly rough surface characterized by small pinnacles of snow, called "nieves penitentes" when they reach appreciable size. Full development of these pinnacles can reach a relief of 2 meters or more, but this latter situation is limited to high altitude glaciers and snowfields, primarily in the tropics. Incipient formation of small nieves penitentes, like those shown here, is fairly common on south-facing slopes of temperate-zone snowfields when the sun altitude is high and the air is dry. In practical terms, this means it is mostly encountered at higher altitudes, lower latitudes, and toward spring. There are two clues here about snow conditions: (1) those favoring nieves penitentes divert an appreciable amount of incoming solar radiation to evaporation, hence extensive melt is limited, and their formation is not usually associated with wet snow avalanches on that particular slope; (2) the rough surface thus generated provides a poor sliding surface for subsequent snowfalls and hence inhibits slab release.

Another surface feature produced by complex heat-transfer phenomena has been termed radiation recrystallization. Figures 31 and 32 illustrate the key features for its recognition in the field. When snow-surface heat balance is negative (cold, dry air and strong long-wave radiation loss cool the snow) while at the same time solar radiation penetrates the near-surface layers to cause melt, an intense temperature gradient exists just below the surface, causing very rapid kinetic growth recrystallization. When the sun goes down, the subsurface melt layer freezes; the net result is to produce a thin layer of very loose, fragile crystals on top of a crust. Unlike surface hoar, which forms from atmospheric water vapor, these crystals are formed from already-deposited snow. Figure 31 shows in an area about 15 cm wide the external appearance revealed when the loose crystals have been swept away by a gloved hand. The firm crust underneath has resisted the sweeping motion and exhibits a distinctly different texture from the loose, recrystallized snow above. Figure 32 shows the structural details in profile. A small pit has been dug into the snow and a bridge has been left suspended across the space where underlying snow was removed from beneath the self-supporting crust, which appears here as a smooth, dark layer about

31 *San Juan Mountains, Colorado*

1 cm thick. On top of the crust is a layer of recrystallized snow with obviously different texture.

Radiation recrystallization may occur on level ground, but it is most common on south exposures in late winter or early spring. It is more common at high altitudes where atmospheric conditions favor the intense long-wave radiation cooling necessary to its formation.

32 *San Juan Mountains, Colorado*

The crust is usually breakable under skis. This phenomenon is a potential source of dangerous avalanches, for the crust provides a good sliding surface for subsequent snow slabs while the recrystallized layer provides the weak bond, or lubricating layer, between slab and crust. This crucial combination, though, can easily be destroyed prior to the next snowfall. Wind may sweep away the loose

33a *Wasatch Mountains, Utah*

crystals, or a shift in radiation balance may destroy the whole structure by melt.

A common new snow feature is the formation of settlement cones around the trunks of deciduous trees (figs. 29, 33a). These usually develop a day or two after an appreciable fall of low-density new snow. As the snow settles during initial stages of metamorphism, the

33b *Wasatch Mountains, Utah*

new layer shrinks in thickness everywhere except where inhibited by local friction of deciduous tree trunks. On conifers, the new snow accumulation often connects with snow canopies on the lower branches. As the new snow settles, a crack can form between snow cover and canopy (fig. 33b). Such signs of settlement are evidence of increasing stability within the new snow.

Avalanche-Related Features

Returning to a wider view of surface features, one of the commonest of all is new snow sluffing on steep slopes, typically above 35 degrees. In figure 34 a benched hillside has produced dry, new snow sluffs on the steep faces of the benches. Most exhibit the typical point origin of loose snow avalanches and start below obvious trees or rock outcrops, which provide falling snow clods as triggers. Extensive sluffing like that seen here indicates a general stabilizing trend in new snow. Caution is still required, though, for soft slabs may have

34 *San Juan Mountains, Colorado*

35 *Wasatch Mountains, Utah*

formed on other slopes, and deep instabilities can still lie buried beneath the innocuous surface sluffs. It helps to know how deep the new snowfall is and what came before it.

The distinction between dry snow and wet snow sluffs is important for evaluating ski and avalanche conditions. Dry snow examples in figure 35 speak of local stabilizing on such a slope and often accompany good powder skiing. The wet snow examples in figure 36 exhibit a different character, with a clear sliding surface and coarser debris accumulation. Occurring early in the day, they may be precursors of larger wet snow avalanches. Ski conditions are often in transition, and skiing quality becomes dependent on timing in respect to solar heating of a given slope.

36 *Brooks Range, Alaska*

37 *Wasatch Mountains, Utah*

A rare pattern of sluffing occurs in figure 37. A poor bond of new snow to a smooth old surface has resulted in a general slumping of the new snow, with characteristic half-slab and half-loose snow. The sign in the foreground indicates the local snow ranger's opinion of this situation. Figure 38 shows additional examples of sluffing in dry snow along the steep ridge to the right. More noteworthy here is the faint evidence of old ski tracks under the blanket of new snow. Soft

38 *Wasatch Mountains, Utah*

slab avalanches in new snow seldom occur on slopes where old ski tracks in the previous snow surface can still be recognized. The old surface broken by ski tracks does not form a good sliding surface and the new snow usually is too shallow to form avalanches internally.

39 *Wasatch Mountains, Utah*

Breakable crust is the skier's bane, but fracturing or cracking of the snow cover is a feature critical for recognizing slab avalanche development. This may be seen as visible cracks, heard as audible settling sounds, and felt as the motion of settling or displacement. In figure 39 the shallow surface crust has cracked and broken away under the skis only where actually disturbed. There is no propagation of fractures as evidence of stresses in the snow, and this situation in itself exhibits a harmless snow condition, though less than

40 *Wasatch Mountains, Utah*

optimum skiing. Figure 40 shows a deeper fracturing layer with some very local extensions of the cracking beyond the disturbing skis. This pattern may not be serious locally, but other slopes with a deeper deposit of the hard surface snow may be approaching a slab condition.

The warning signs are getting clear in figure 41, where the weight of a skier near the crest of a cornice has sent a crack propagating some distance ahead. This is a risky route to follow simply from the prospect of cornice collapse, while the cracks warn that similar snow

41 *Wasatch Mountains, Utah*

stresses may be found in avalanche release zones. This latter possibility is realized in figure 42, where cracking in the foreground snowdrift has propagated all the way onto the steep lee slope beyond (arrow indicates the latter development). Finally, the phenomenon

42 *Wasatch Mountains, Utah*

reaches its full development in figure 43, where skier disturbance at a ridge crest, as in figure 42, here has propagated into a slope and actually released a slab avalanche whose crown face is visible at the end of the cracks. Dangerous conditions are signalled by such cracking, especially when accompanied by cracking or settling

43 *Wasatch Mountains, Utah*

sounds. This situation demands great caution by travelers in avalanche terrain. When such cracks run for 10 meters or more (a kilometer is possible), the instability has become very serious indeed. The importance of this visual evidence cannot be too strongly emphasized: *Beware of snow that supports fracture propagation.*

44 *Parsenn resort area, Davos, Switzerland*

A different but related phenomenon is called a glide crack (fig. 44). The snow cover slides (glides) slowly over the ground surface throughout the winter wherever the snow remains near the melting point and the ground is smooth enough to allow it. In favorable circumstances this gliding becomes active enough to make the snow cover break into cracks at points of stress. Such gliding is most vig-

45 *Mount Shiribetsu, Hokkaido*

orous in winters that started with deep falls of snow in the autumn when the ground was still warm, for this enables a heat supply to ensure melting at the snow cover bottom. Glide cracks are not always local; they can spread clear across a mountainside (fig. 45). Where the ground surface is unusually slippery (mountain bamboo in figure 46) and snow conditions favorable, once cracks form, the glide motion can accelerate over a period of several hours until the motion becomes an avalanche. In Japan where such events are common, these are called transavalanches.

46 *Toikanbetsu, Hokkaido*

Snow in Trees

The way that snow lies (or does not lie) in trees contains many clues to the character of recent snowfall and weather. These clues add to the store of visual snow information available to the alert observer.

When predominantly stellar crystals fall in light wind, the deposited snow exhibits canopying as the crystals interlock during accumulation. Cushions of snow begin to build up on exposed objects, notably including tree branches, until the enveloping snow grows to a much larger diameter than the object it covers. Moderate canopying is seen in figure 47, evidence that a limited amount of snow has fallen, probably less than 30 cm. With each branch still freshly covered with snow cushions in bright sunshine, the observer can recognize that the snowfall has just ended and deduce that the air temperature is far enough below freezing to inhibit any immediate melting. Only a moderate snowfall is suggested because not enough snow has built up on the cushions to start sliding away, and the cushions remain fat and rounded.

Figure 48, on the other hand, shows the canopying associated with a deep snowfall. Here the weight of the accumulated snow has begun to bend the branches downward, and the snow has piled up deep enough to start shedding loose sluffs from the cushions, so that many of the snow cushions on the branches have lost their rounded character and now have a wedge-shaped appearance.

Once the snow is deposited in cushions, metamorphism immediately begins to break up the stellar crystals and weaken the interlocking bonds. When this happens the snow loses cohesion and starts falling from the trees (fig. 49). As a lump of snow descends, it often

47 *Wasatch Mountains, Utah*

48 *Cariboo Mountains, British Columbia*

49 *Bugaboo Mountains, British Columbia*

sweeps other lumps along with it until a regular cascade of snow arrives at the bottom, pocketing the surrounding snow with impact craters like those in the foreground of fig. 50. Freshly fallen in new snow, these "tree bombs" offer little resistance to skiing. Once age-hardened for a few hours, they become tough, resisting lumps lurk-

50 *Bugaboo Mountains, British Columbia*

ing under the next snow accumulation. Falling lumps and tree cascades can easily serve as avalanche triggers if they occur in a release zone. Below timberline, the post-snowfall period of initial metamorphism and snow dumping from trees raises the danger level in an unstable snow cover.

51 *Bugaboo Mountains, British Columbia*

External forces, such as wind, also cause trees to shed snow. When a strong wind gust strikes a snow-laden forest, large amounts of snow are dumped from trees as well as carried into the air as dust clouds (fig. 51). When this happens, avalanche triggers abound, and avalanche release in unstable snow becomes highly probable. Gusty winds following deep new snow should always be regarded with caution, especially in forested terrain. Even modest amounts of heat on a sunny day can also start the shedding process. In figure 52 this process is underway, with slender branches already showing gaps in the festoons of new snow, and the pocketed appearance of the sur-

52 *Mount Teine, Hokkaido*

face below showing where the departed snow has gone. When strong solar heating, common in late winter and spring, follows a new snowfall, it can rapidly cause melt instability in the deposited snow and at the same time generate many avalanche triggers as trees shed their loads.

Not all snow accumulation in trees comes about through canopying as snow crystals interlock. Near the freezing point snow becomes wet and sticky, so that it readily adheres to every branch and twig and then may freeze in place. In this case snow sticks to each separate branch and twig without building wide cushions (fig. 53).

53 *Cascade Mountains, Washington*

54 *Wasatch Mountains, Utah*

Freezing rain occurs when liquid droplets fall through a layer of air colder than the freezing point and become supercooled. These supercooled rain droplets, very large compared with the tiny supercooled cloud droplets that form rime, spread out and freeze to form a layer of clear ice when they strike any kind of surface. This phenomenon is sometimes called a silver thaw. When coating trees, as in figure 54, it rapidly adds weight to cause bending or collapse. Coating road surfaces, it leads to traffic disasters. When freezing rain falls it also coats the snow surface with a hard, slick crust.

55 *Japanese Alps, Honshu*

Depending on its thickness, this can lead skiers into breakable crust or onto hard, smooth ice, making dangerous falls possible on steep slopes. This smooth ice crust makes an excellent sliding surface for avalanches generated by subsequent snowfalls.

Quite a different accumulation comes from rime, or the freezing of tiny supercooled liquid cloud droplets on exposed objects. If rime alone is present, its arrangement on trees is shaped by the direction of the wind bringing the supply of cloud droplets. Rime builds toward this supply and hence lies on the windward side of trunks and branches. If snowfall accompanies rime formation, the rime serves to cement the snow crystals together when they strike trees or other exposed objects. Both of these phenomena appear together in figure 55. At one time, a rime-laden wind from the left has plastered the accompanying snowfall against the left side of these birches. At another time (the sequence is not clear from the picture), wind bearing rime alone in the absence of snowfall has blown uphill from the right and formed a feathery rime accumulation on the right side of the trunks. A combination like this sets the observer wondering what the effects have been on the snow cover, but there is no clear answer here without other lines of evidence. (An extreme example of rime-cemented snow on rocks appears in figure 1, p. 6.)

When riming takes place on trees, it also occurs on individual snow crystals as they fall through the air. Although rime-cemented snow on windward exposures tends to be stable, this definitely is not true in more protected accumulation zones, where rimed snow crystals often lend a quasi-stability to soft slabs that allows them to build up to a dangerous thickness before avalanching. The role of snow-crystal riming in the quality of deep powder skiing has already been mentioned.

A whole winter's snow history is encompassed by the tree snow-cushion in figure 56, where springtime melt has long since cleared most tree-borne snow. Faint evidence of layering speaks of accumulation by several storms. The dimpled appearance of the surface shows that some melt by both sun and warm wind has occurred under the same conditions that tend to form sun cups on melting snow and firn. Icicles hanging from the bottom tell of sun-driven melt during subfreezing air temperatures. Old accumulations like this are dense and heavy; they fall as dangerous "tree bombs" when they eventually do come loose.

56 *Teton Mountains, Wyoming*

Melting Snow

Melting snow offers another whole spectrum of surface features serving as clues to snow behavior and stability. These clues are especially important in the sometimes rapid transition from a subfreezing snow cover or new snowfall to a melting state with free water present.

The sequence of sunball formation provides easily recognizable clues to fast-changing ski conditions and the evolution of wet snow avalanches. Sunballs are rolling lumps of snow that start as clods falling from trees, cornices, rocks, or the passage of skis. They are properly called sunballs because their downhill progress and growth depends on subsurface melt in snow that comes about from solar radiation. This phenomenon is less common in snow wet by rain, except when deep wetting may produce large balls or wheels. Figure 57 shows the earliest stage of sunball formation. A fall of new snow is exposed to the sun late in winter, when the near-surface layers immediately absorb transmitted solar radiation and melt begins. As the first traces of liquid water appear, the snow becomes slightly sticky. Good deep powder skiing may still prevail at this point, for the bulk of the new snow is predominantly dry, but dislodged lumps of snow can roll and even grow, as they have in this photo. This is not yet the stage where wet snow avalanching begins, but it can signal the early onset of natural releases in unstable soft slabs.

In figure 58, melt has progressed to the next stage. Appreciable liquid water now appears in the top few centimeters of snow. Even very small snow lumps roll downhill and accrete wet snow. Avalanching at this point is usually limited to very thin surface layers

57 *Wasatch Mountains, Utah*

58 *Wasatch Mountains, Utah*

on high-angle slopes. Skiing quality definitely takes a turn for the worse.

Before going on to subsequent melt stages, it will be useful to consider a less frequently observed version of the stage in figure 58. This is illustrated in figure 59, where close inspection of the shadows cast by the sunball tracks reveals that these stand as ridges above the surrounding snow surface instead of being formed as grooves depressed below the surface. In this case an earlier sunball episode much like that of figure 58 has terminated at that stage. Subsequent weather has allowed the snow to settle and stabilize without additional melt. Like the ski tracks in figure 9 (p. 18), the snow initially compressed as grooves by the rolling sunballs has hardened and failed to follow the settlement of the surrounding snow, ending up as ridges. The sunball tracks in figure 59 may be as much as several days old and no longer speak of current snow conditions. The settlement suggests that any wet snow avalanche hazard may have diminished by the time this photo was taken.

As solar heating continues on south exposures during a clear day, more and more random lumps of snow fall from rocks and trees to find an increasingly wet snow surface on which to roll. The transition from conditions illustrated in figure 58 to those in figure 60 is often rapid. The sunballs increase in number and reach larger sizes as they descend over wet snow. Serious wet snow avalanche danger is still quite limited, but if this transition has been rapid it signals a clear warning that snow stability may also rapidly deteriorate in the next hour or two. Such deterioration is already approaching in figure 61, where significant free water has penetrated 30 cm or more, and the sunballs are starting to turn into larger snow wheels. A similar situation appears in figure 62, where the deepening penetration of melt-water in the snow allows the highly deformable wet snow to form incipient wheels and deep grooves. When the snow reaches this stage, wet snow avalanches are not far behind. Figure 63 is a good exam-

59 *Wasatch Mountains, Utah*

60 *Wasatch Mountains, Utah*

ple of the avalanches, along with a fine specimen of a snow wheel a full meter in diameter.

To sum up the nature of these clues to developing wet avalanche conditions, small sunballs are seldom concurrent with big avalanches unless an unstable slab condition exists for other reasons, but they point the way to changes to come. When large sunballs evolve into snow wheels, the danger point for wet snow avalanches clearly has been reached. Depending on weather conditions and time of year,

61 *Wasatch Mountains, Utah*

62 *Wasatch Mountains, Utah*

63 *Wasatch Mountains, Utah*

this transition can happen very rapidly, often within an hour or two. Wet loose snow avalanches are the usual result. Sometimes large wet slabs may fall, but this often follows some loose snow activity.

Turning from sunball evolution to other wet snow features, consider the way the snow lies on the landscape in figure 64. Along the streambed in the foreground and middle distance, the snow cover exhibits a wrinkled pattern around local holes and bumps, suggesting that the whole depth of snow has sagged and settled. This is exactly what has happened as a spring thaw or the melting induced by the buried stream has warmed a cold snow cover to the melting point and initiated a sudden increase in settlement. The constraints of surface geometry require this settlement to be absorbed as the sagging and wrinkling seen here. The key conclusion is that the snow has already made the critical transition from subfreezing to an isother-

64 *Selkirk Mountains, British Columbia*

mal temperature regime, and a likely peak of avalanche activity has already passed. There are several other clues in this photo as well. In the upper right there has been a recent and vigorous activity by small sunballs, indicating current surface melting. Just above left center, the larger debris of an earlier wet snow avalanche is visible, its surface partially obscured by a more recent and light fall of new snow, probably the same fall that set the stage for the sunball activity. The trees in the upper left exhibit the characteristic "flag" effect of small conifers in an active avalanche path, their uphill branches stripped away but their downhill branches intact. Many of these trees also tilt sharply downhill at varying angles, showing the effect of weight from earlier deposits of avalanche debris. The distant skiers and their tracks at top center suggest that good spring skiing conditions exist, with surface layers of a frozen snow cover softened by solar radiation.

The circumstances in figure 65 are subtler. The prominent cornice brow in the right foreground shows a textured surface associated with appreciable weathering and melt. The same is true, though less distinctly seen, on the face of the distant cornice. The latter, together with the slope immediately below, has an unusually bright gleam associated with incipient firnspiegel formation, indicating a combination of subsurface solar melting and surface freezing due to dry, cold air. The snow surface in the foreground and middle distance is smooth and unweathered, unlike the cornice faces. The large sunballs dislodged by passing skis suggest a rather high free water content in the top snow layers. Given the observer's knowledge that the season is late March, a picture emerges of an old and settled snow cover with no recent episodes of snow drift and cornice building, while a very recent snowfall with little wind has left the smooth snow surface. Cold, dry air and the accompanying long-wave radiation cooling has slowed or prevented surface melting on south exposures, while subsurface melt proceeds. Ski conditions are variable with

65 *Teton Mountains, Wyoming*

66 *Wasatch Mountains, Utah*

exposure, and probably less than ideal. Depending on what the earlier old snow surface was like, some wet snow avalanching in the new snow could be just ahead.

Rain is a notoriously effective trigger for avalanches when it falls on new snow, or even on a mature snow cover that is at subfreezing temperatures. The reaction of the snow is very fast in these cases. In either case, the introduction of abundant liquid water into the snow leaves behind an aftermath of distinct surface patterns. Usually by the time such features are developed and recognized, the snow has become stable, either by extensive settlement or by refreezing of the liquid water in the surface layers to provide a very strong bond among snow grains. Rainwater does not percolate into snow

67 *Sapporo, Hokkaido*

by even diffusion, but instead concentrates very quickly into subsurface flow channels. Settlement is locally accelerated around these channels. This settlement is soon reflected at the surface by depressions or rills that follow the course of the subsurface channels. The effect varies in degree according to the amount of liquid water and the original character of the dry snow. Figure 66 shows a striking example of this pattern on a sloping hillside, where surface rills clearly locate each buried percolation channel. These features are also visible in figure 2 (p. 6). On level ground the subsurface channels tend to form vertically in the snow cover, leading to a dimpled surface like that in figure 67. Both examples illustrate snow conditions well after rain has done its work. When deep, the surface

68 *Wasatch Mountains, Utah*

grooves can make for rough skiing, especially if the wet snow has frozen hard.

When the surface layers of new snow acquire a certain amount of free water, they pass rapidly through states of plasticity that are difficult to describe in print but which are readily noted by a skier's kinesthetic sense working through boots and skis. One of these intermediate states turns sideslipping or stemming of skis into a stick-slip process that causes awkward chattering of skis. Figure 68 shows the visual expression of this stick-slip process in the corrugated surface of the ski tracks. When conditions like these appear, the snow is changing fast as increasing amounts of meltwater appear. The snow stability may also be changing fast, and the alert observer seeks a wider circle of clues about what will come next.

Other Snow Features

Many other related topics besides visual snow features, such as snow stratigraphy, mechanical properties, thermodynamics, and acoustic behavior, contribute to interpreting snow conditions. Although a full treatment of these topics is far beyond the scope of this book, a few lesser-known sidelights are introduced below.

The discussions to this point have made several references to the kinesthetic sense, the "feel" of the snow as it reacts beneath skis. In this respect there is a curious feature to the stick-slip process illustrated in figure 68. I once attached a microphone to one of my skis and connected it to a tape recorder. Skiing in a wide variety of snow conditions produced a collection of characteristic snow sounds that varied with the type and metamorphosed state of snow crystals. One of these sounds was a clear record of ski chatter while making a turn in snow like that of figure 68. Sound analysis plots of this particular record are shown in figure 69, where the chatter interval is labelled with a bracket. The lower plot shows waveform, or amplitude of the total sound as a function of time. The upper plot is a spectrum analysis, where frequency is shown as a function of the same time interval in seconds, with intensity of the sound at different frequencies depicted by darkness of the plot. The physically perceived ski chatter clearly was recorded in the waveform, but the loss of higher spectrum frequencies during chatter was an unexpected result. The point of this discussion is what came in the next test. The ski-microphone combination was sensitive enough to pick up weak chatter even when there was no perceived sensation through skis, boots, and feet. This raises the question of how much of the "feel"

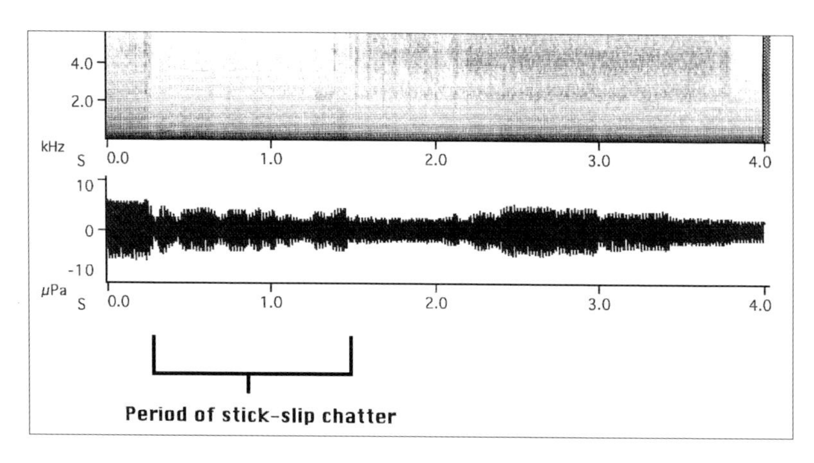

69 *Period of stick-slip chatter*

of snow is a subconscious perception of similar vibrations. The mechanical characteristics of the skis obviously must play a part. There are some opportunities here for interesting research.

Not all snow features are exhibited at the surface. In figure 70 a rotary snow plow has exposed from ground to surface a spring cross-section of the snow cover along a roadside. A signpost in the middle has locally inhibited snow settlement during winter accumulation. The highly visible layering in the snow brings out the fact that settlement is inhibited not just above the post but in diminishing amount over a distance ten times or more the diameter of the post. The post is not just supporting the snow immediately above it, but a much wider, and thus heavier, body of snow. This is why snow settlement can exert very large forces on buried structures. Consider the example of a cabin in the deep snows of a maritime climate. As winter snow accumulates, it can pile up deeper and deeper on the roof, and the cabin needs to have the structural strength to support this snow load. If the surrounding snow cover on the ground piles

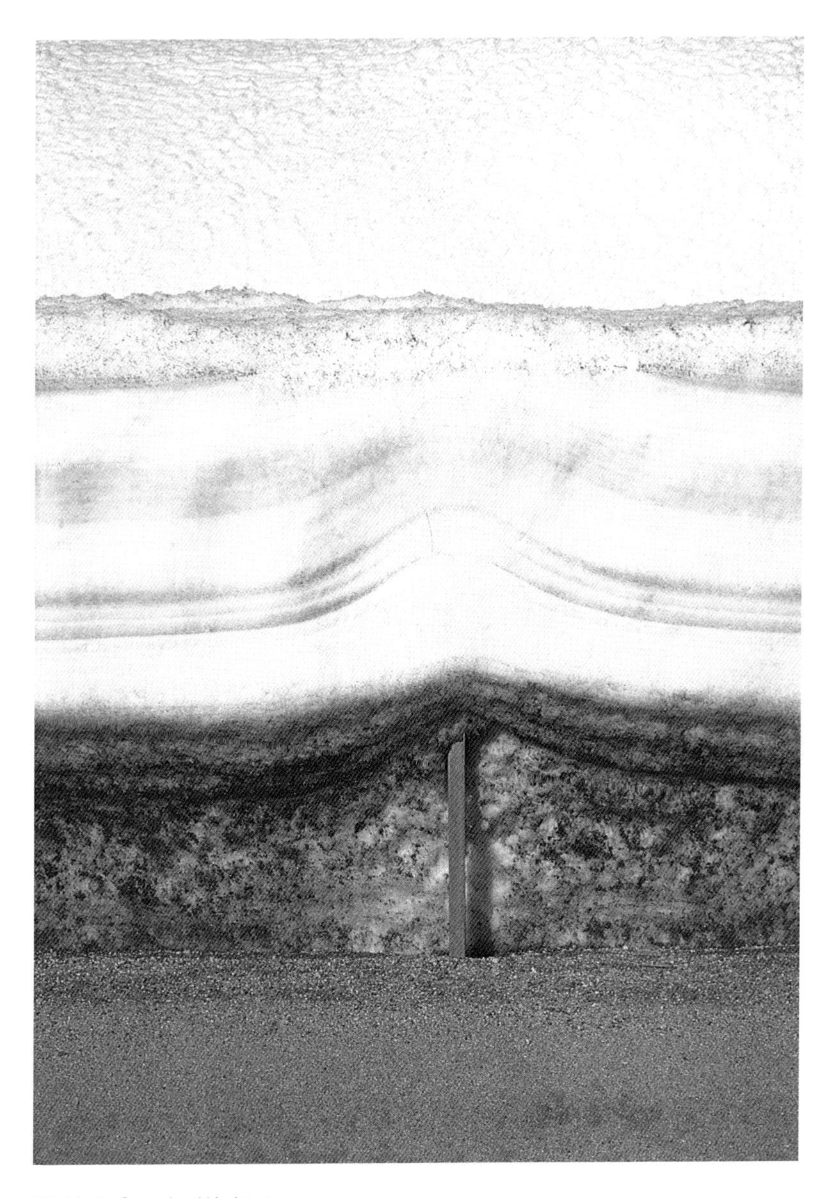

70 *North Cascades, Washington*

up deep enough to connect with the roof snow, then the situation changes radically. Now the cabin has to support not only the roof load but also the downward pull of settling snow for a large surrounding area. Most structures are not designed for these kinds of forces and sometimes will collapse.

The layers visible in the upper half of the snow in figure 70 stand out because melt has caused dirt particles preferentially to collect on variations in grain size or on ice layers. There are deliberate ways to bring out this stratigraphy for inspection. The variations in grain size and type can be developed like a photographic film by spraying the snow-pit wall with a water-soluble dye (fig. 71). Quickly passing a torch flame over the dyed surface enhances the effect. An unplanned version of this appears in figure 72, where heat and smoke from a bonfire built next to a snow face have brought out the layered structure of the snow. The variations in hardness can be delineated by stroking the pit wall with a soft brush, as in figure 73.

Reference was made earlier to the phenomenon of age-hardening in snow, where disturbed snow gains strength with time owing to new bonds growing among the grains. Figure 74 illustrates an unusual example of this. A shallow snow cover consisting entirely of fragile, almost cohesionless depth hoar was compacted under snowshoes and allowed to stand for 24 hours at temperatures well below freezing. At the end of that time, age-hardening had done its work, and the compacted snow could be cut into blocks and even stacked to form a wall.

Roof avalanches are a common winter feature in snow country. They can be dangerous, often involving many tons of snow. Attention to their precursors, such as glide cracks, spring thaw, or heating of a previously cold building, is a good safety practice. Roof avalanches sometime develop as transavalanches like those in figure 46 (p. 58), starting first with glide cracks and then accelerating into avalanches. Figure 75 is a striking example of snow glide on a roof, with an

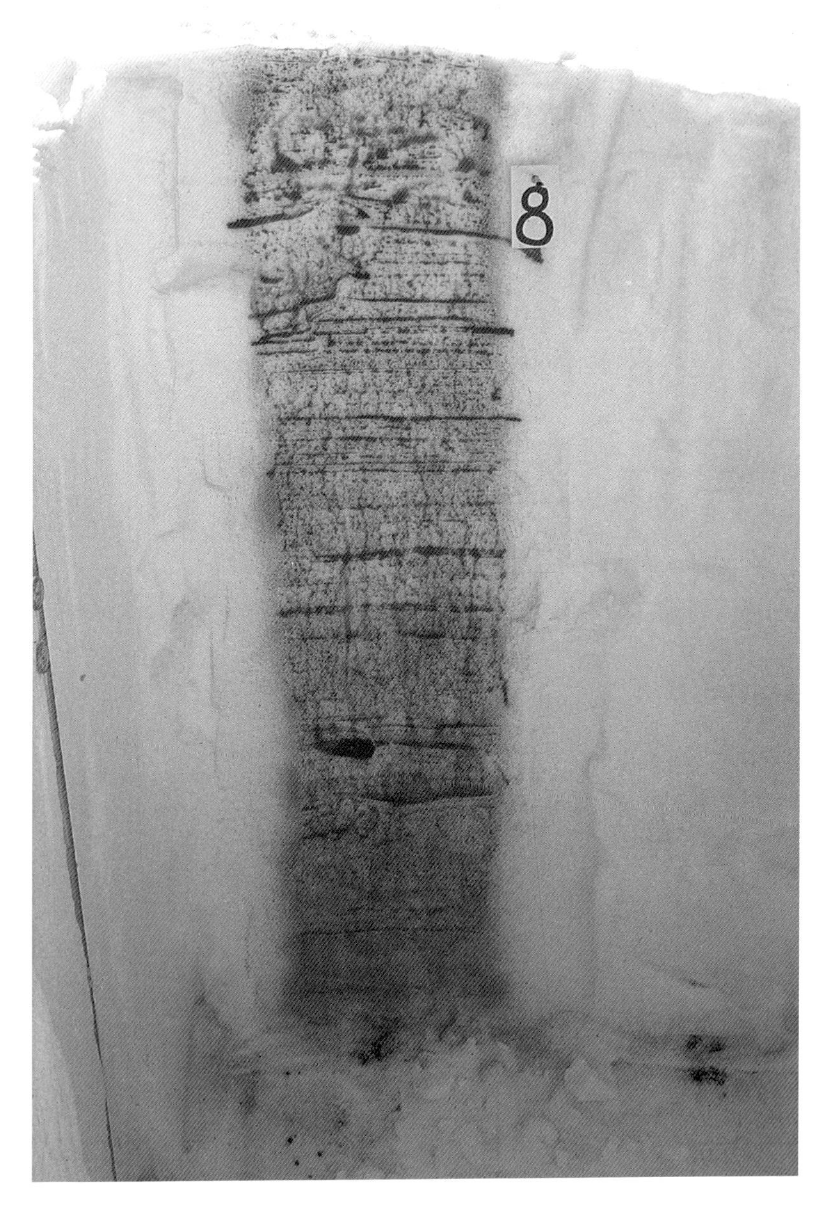

71 *Cascade Mountains, Washington*

72 *Wasatch Mountains, Utah*

73 *Bugaboo Mountains, British Columbia*

74 *Wrangell Mountains, Alaska*

75 *Cascade Mountains, Washington*

avalanche about to follow. Note that the snow has a low enough viscosity while still being cohesive, so it can bend without breaking. This kind of deformation forms the curl of cornices like those in figures 14 and 15 (pp. 23 and 24).

Metal roofs, as in figure 76, readily promote sliding, composition roofs less so. Stepped roof surfaces like cedar shakes or tiles also avalanche readily because any snow glide steadily decouples the snow from contact with the roof and its accompanying friction. If a whole winter accumulation has to be removed from a roof, there are a couple of ways to accelerate removal by avalanching. One is to dig a trench along the roof peak, thus dividing the snow layers on both sides, allowing them to break free from each other and slide more easily. Another is by thermal cycling. Heating the building often will promote a roof avalanche by melting snow next to the roof surface, which is especially effective with metal. If this doesn't work, shut off the heat and let the building cool below freezing, so the water-saturated layer of snow next to the roof freezes into an ice layer. Then reheat the building. Wet ice slides much more easily than wet snow.

The building with the fresh roof avalanche in figure 76 is located in the center of figure 77. I once watched a roof avalanche (although not the one pictured in figure 76) fall from this building. Within seconds most of the scattered roofs in figure 77 also avalanched, giving credence to the debated hypothesis that avalanches sometimes can be triggered by infrasound (very low frequency vibrations). As with the stick-slip sounds or the audible cracking described earlier, the "voice" of the snow may have a lot to tell us if we learn how to listen.

All of our senses are busy collecting information about the world around us. In the everyday world this deluge of data for the most part goes unused. Out on a mountain slope on a snowfield, perhaps with a storm in progress, with avalanche dangers all around, the

76 *Wrangell Mountains, Alaska*

stress of survival takes over. Every bit of information becomes important, but we tend to emphasize with our attention those bits that are easy to remember, to write down, perhaps to put numbers on, because that is the way we have learned to assimilate and record information about the world. All the rest of the information, less easy to define or describe, also keeps flowing through our senses and influences our decisions. The "whumpf" sound of a settling snow slab grabs our attention with electrifying effect because survival is at stake. The character of rattling when wind drives snow against a parka hood speaks of crystal types and hence the nature of accumulating lee drifts. The kinesthetic sense of skis running through a layer of new snow tells the experienced mountain guide whether the snow is likely to avalanche. These kinds of data are just as important as measurable information like air temperature or wind velocity or slope angle, or visual information like the curve of a cornice. They form just as important a part of interpreting snow conditions as the thickness or density of a new snow layer, but are relegated to the realm of seat-of-the-pants feel or intuition because they are

77 *Wrangell Mountains, Alaska*

difficult to communicate from one person to another. They are even more difficult to put into a book.

This book has dealt with the visual aspects of snow on the mountain landscape. This is the obvious sequel to the microscopic world of *Field Guide to Snow Crystals*, because the visual data are the easiest to collect. Even so, it is important to reiterate the importance of attention. Even the most conspicuous snow features offer little information if the mind's eye is not working together with the physical eyes. We see what we know, but overlook the unfamiliar. The review of snow features found here will serve its purpose if readers enlarge their *knowing*, the mind's eye at work recording and interpreting the many details of the snow surface.

Glossary/Index

Text page references follow the definitions. Numerals in parentheses indicate figures where features are displayed.

Ablation: The loss of mass from snow or ice, most commonly by melt but also by sublimation or scour. 35

Age-hardening: A characteristic property of disturbed snow to gain strength with time through bond growth among the rearranged snow particles. 15, 16, 89 (7, 9, 74)

Avalanche, loose snow: A sliding mass of snow that originates at a point on a slope as a small surface clump of snow and sweeps up larger and larger snow quantities as it falls. 31, 46, 79 (34, 35, 36, 38)

Avalanche, slab: A sliding mass of snow that breaks away as a discrete, cohesive layer from the clearly defined line of failure (the fracture line, or crown face). 13, 16, 22, 37, 51 (cover photo, 43)

Condensation: The phase change from gas (water vapor) to liquid (water) or to solid (ice). 20, 27, 35

Cornice: An overhanging mass of snow deposited on the lee of a ridge or other obstacle to drifting snow. 23, 24 (14, 15, 16)

Depth hoar: Fragile crystal formed within the snow cover when strong temperature gradients induce existing snow crystals or grains to recrystallize by passing through the vapor phase. 9, 16, 18, 20, 89 (74)

Evaporation: The phase change from liquid (water) to gas (water vapor). 40

Failure plane: The surface between two usually cohesive snow layers (slabs) and often the sliding surface of a slab avalanche. 21, 22

Firn: Snow, usually on a glacier, that has survived a complete season of ablation. 27, 35, 69 (20)

Firnspiegel: A thin, usually fragile layer of smooth, clear ice that forms over a melting snow surface with just the right heat-balance conditions. Sometimes called "glacier fire" when it reflects sunlight like a mirror. 80 (65)

Hard depth hoar: Dense, fine-grained old snow cemented by tiny vapor-deposited crystals between the grains. Like depth hoar, it is generated by strong temperature gradients, but results in a gain instead of loss of strength. 18, 20 (11)

Kinetic growth: Metamorphism of snow crystals by recrystallization under the influence of strong temperature gradients. 41

Metamorphism: Changes in size and shape of snow crystals or grains governed by the internal environment of the snow cover. 44, 59

Radiation cooling: Loss of heat from a snow surface when the radiation heat balance is negative for that surface. Loss of energy in the long-wave (infrared) part of the spectrum dominates this kind of cooling. It is especially strong with dry air, clear skies, and high altitudes. 8, 42, 80

Release zone: The part of a slope, usually the upper part, where an avalanche breaks loose. This may apply to both loose snow and slab avalanches. 63 (cover photo, 4, 16, 34, 36, 37, 38, 43, 44, 45, 46, 63, 77)

Rime: A dense, fine-grained deposit built up from the freezing of supercooled cloud droplets onto exposed objects, including falling snow crystals. 5, 8, 33, 38, 39, 40, 68, 69 (frontispiece, 1, 28, 55)

Snow climate: The characteristic combination for a given area of winter weather parameters, such as temperature, wind, snowfall amounts, and percentage of clear skies. 5, 8, 9, 16 (1, 2, 3, 4)

Snow creep: The internal deformation of the snow cover under the influence of gravity resulting in downhill displacement on a slope.

Snow crystal: An individual particle of snow, often with some form of hexagonal symmetry, whose water molecules are arranged in a common crystallographic orientation. 69

Snow density: The weight of a snow mass per unit volume. Very low density snow "fluff" may weigh 20 to 40 kilograms per cubic meter. Typical new snow densities are 70 to 120 kg/m^3, while dense old snow may reach 400 to 500 and firn 600 kg/m^3. The density of bubble-free ice is 917 kg/m^3. 8, 11, 13, 16, 18, 26, 30, 33, 44, 94

Snowflake: An aggregation of individual snow crystals. 30

Snow glide: The slow sliding of the snow cover along the ground under influence of gravity on a slope. The amount of glide is highly dependent on ground roughness and character. It is not the same as snow creep, defined above. The total displacement of the snow surface on a slope is the sum of creep, glide, and settlement. 56, 89 (44, 45, 75)

Snow settlement: Compaction of the snow cover through metamorphism and compression. On a sloping surface, additional deformations occur from snow creep and snow glide, defined above. 44, 45, 74, 79, 82, 87 (33a and b, 70)

Sublimation: The phase change from solid (ice) directly to gas (water vapor) at subfreezing temperatures without an intermediate liquid (water) phase. Sublimation sometimes also refers to the reverse (water vapor directly to ice) but is not used here in this latter sense. 11, 16

Supercooled: Water that remains liquid when cooled below the freezing point is said to be supercooled. Tiny droplets, such as those that make up clouds, are much more easily supercooled than larger masses of water. 5, 38, 68

Surface heat balance: The algebraic sum of heat inputs (plus) and losses (minus) to a surface. For snow, these include sensible heat by conduction from below or turbulent transfer from the atmosphere, condensation or evaporation, and both long (infrared) and short (visible) radiation. The snow surface heat balance can be either positive (gains heat) or negative (cools) according to circumstance. 41

Wind scoop: A hollow space formed in a snowdrift on the windward side of an obstacle. 15 (7)

Bibliography

Avalanche Atlas: Illustrated International Avalanche Classification, UNESCO. Paris: UNESCO, 1981.

The current standard for morphological and genetic classification of snow avalanches. Illustrated with a superb collection of photographs, not only of avalanche features in great detail, but also sections on snow crystal types, snow cover stratigraphy, and many unusual surface features. Well worth studying for the many visual snow features displayed and their relation to avalanches. Parallel texts in English, German, French, Spanish, and Russian.

Doeskin, Nolan J., and Arthur Judson. *The Snow Booklet: A Guide to the Science, Climatology and Measurement of Snow in the United States*. Fort Collins, Colo.: Colorado State University Department of Atmospheric Sciences, 1996 and 1997.

Characteristics, history, statistics, anecdotes, and measurement methods of snow. A wealth of information in a small and nicely illustrated package.

Fredston, Jill, and Doug Fesler. *Snow Sense: A Guide to Evaluating Snow Avalanche Hazard*. Anchorage, Alaska: Alaska Mountain Safety Center, 1994.

Very well-illustrated, practical information on the art of paying attention to the kind of snow conditions that lead to avalanches, and hence to the character of snow in general.

Hoeg, Peter. *Smilla's Sense of Snow*. New York: Farrar, Straus & Giroux, 1993.

A novel—actually a murder mystery—with a plot revolving

around interpretation of snow clues. A film version is available in video stores.

International Classification of Seasonal Snow on the Ground. International Commission of Snow and Ice, International Association of Hydrologic Sciences, 1990.

Current scientific thinking on snow crystal metamorphism and terminology, with both genetic and morphological classifications. Updates material in *Field Guide to Snow Crystals.*

LaChapelle, Edward R. *Field Guide to Snow Crystals.* Seattle: University of Washington Press, 1969.

Companion volume to *Secrets of the Snow.* Diagrams and photomicrographs of snow crystals as they appear in nature, with explanations of the physical processes that form and alter them.

McClung, David, and Peter Schaerer. *The Avalanche Handbook.* Seattle: The Mountaineers Books, 1993.

This is the most recent version of the basic North American manual for snow avalanches. It contains extensive treatment of the snow cover and meteorological influences upon it, much of this relevant to interpreting visual character of the snow surface, with special reference to avalanche formation.

Seligman, G. *Snow Structure and Ski Fields.* London: Macmillan, 1936.

Available in reprint edition from International Glaciological Society, Lensfield Road, Cambridge CB2 1ER, UK. Although in many respects scientifically dated, this classic treatise provides a wealth of material on formation, structure, and interpretation of snow features.

Library of Congress Cataloging-in-Publication Data

LaChapelle, Edward R.
Secrets of the snow : visual clues to avalanche and ski conditions /
Edward R. LaChapelle
p. cm.
Includes bibliographical references and index.
ISBN 0-295-98151-2 (alk. paper)
1. Avalanches—Control. I. Title.
TA714.L34 2001
551.57'84'0247969—dc21 2001035592

National Library of Canada Cataloguing in Publication Data

LaChapelle, Edward R., 1926–
Secrets of the snow
ISBN 1-55054-884-0
1. Snow. 2. Avalanches. 3. Skis and skiing—Safety measures. I. Title
TA714.L32 2001 551.57'84'0247969 C2001-910905-9

Greystone Books gratefully acknowledges the support of the Canada Council for the Arts and the British Columbia Ministry of Tourism, Small Business and Culture, as well as the financial support of the Government of Canada through the Book Publishing Industry Development Program (BPIDP) for its publishing activities.